Rocket Rangers

Man's Quest to Fly Like Buck Rogers

Aerospace Heroes Commemorative

Vol. 1

With vivid photo histories of the
Jetvest, Rocketbelt, Jetbelt, MMU, SPK and S.A.F.E.R

Edited by:

Nelson Louis Olivo

Aerospace Historian

Library of Congress Control Number: 2011906845

ISBN: Softcover 978-1-4628-6546-8
Hardcover 978-1-4628-6547-5

This book was printed in the United States of America.

To Order More Books
The Manhattan Literary Agency
1-718-403-0256
the Xlibris Bookstore at www.Xlibris.com
Amazon.com
Barnesandnoble.com

Summary: A vivid photo history of man in free flight from the days
of Buck Rogers and the early rocketbelts to the era of the MMU.

For

Margarita Pagan – my Mom

And in loving memory of

Judy Wells, Ariane Orenstein, Tom Lennon, Thomas Moore,
Gordon Yaeger, Wendell Moore, Robert Courter,
Harold Graham

Dedicated to

the Young Rocket Rangers

the next generation of dreamers and doers

Indiana's Young Rocket Rangers

Front Row: Dawson Pierce, Sierra Crim, Iris Gramajo, Michael Crim
Back row: Jessica Crim, Karen Gramajo, Diana Gramajo

The dream of every earthbound Young Rocket Ranger

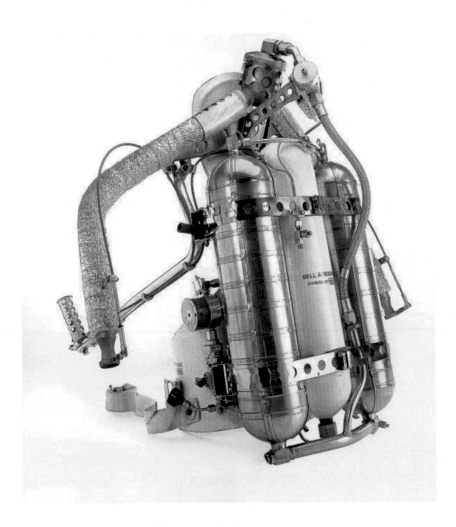

to Fly The Rocketbelt.

Contents Volume 1

Foreword by Physicist Thomas M. Moore
End Page by Rocket Ranger Harold "Pete" Graham
Introduction by Aerospace Historian Nelson Louis Olivo

Foreword

Early man walked on earth and extended his travels with the aid of a wheel. The propeller extended his range in water and especially in the atmosphere. The rocket has freed man from the chains of earth allowing for voyages beyond the moon to Mars, to asteroids and to the vicinity of stars other than our own. This book will familiarize readers with the history with man's quest to apply a rocket to individual flight.

Truly, a great book to begin a countdown to your own rocket-propelled career.

Thomas M. Moore
Physicist and Jetvest Inventor
US Space Camp

FICTION FORECASTS FACT: *Buck Rogers comic strip uses the term "astronaut" in this panel drawn in 1936. The adventure series heralded the Space Age with rocket belts, interplanetary travel, and space stations.*

Before there were space shuttles and spacewalks, landings and moonwalks, Star Trek and Star Wars there was Buck Rogers in the 25th Century.

Science fiction's very first "astronaut" appeared in a story by Philip Nowlan, titled "Armageddon 2419 A.D.," for *Amazing Stories* comics on January 7th, 1929.

Anthony "Buck" Rogers was a former World War I fighter pilot whose peace- time job was to inspect abandoned coalmines for radioactive gases. During one inspection a gas-filled mine collapses. Trapped, Buck loses consciousness, but instead of killing him, the gases preserve him in a deep sleep. When he wakes and escapes, he finds himself in the 25th Century, where aero planes are a thing of the past and people fly through air and space with rocketpacks.

The Adventures of Buck Rogers in the 25th Century, which gave its readers a view into the future, became an instant sensation. A weekly Sunday comic strip followed that was published in over 400 newspapers and translated into18 languages – Buck was read around the world! The popularity of the futuristic series led to a radio program (1932-1947) listened to by legions of fans and astronaut hopefuls who came to be known as Buck's Rocket Rangers.

Buck's Rocket  Rangers pin

In 1939, Buck Rogers appeared on the silver screen in his own movie serial. Both young and old alike stood in lines for hours to buy a movie ticket! Fueled by Buck's futuristic adventures, many of those young fans dreamt that one-day rocketpacks would fly them up through the clouds and towards the stars. How could those youngsters know that their favorite science fiction comic strip was forecasting fact and that within a short span of 25 years, aeronautical engineers would begin to seriously look at the impossible and dream those very same dreams?

If you've ever dreamt of flying solo through the air, rocket-riding over grayish lunar craters or simply taking spacewalks 'round our spinning big blue Earth, then come and join the Rocket Rangers; make the reading of this book your take-off point to aerospace careers.

The Jetvest

The first American design for an Astronaut Maneuvering Unit, with recollections from physicist and inventor Thomas M. Moore (as told to Aerospace Engineer Mark Wells)

In 1961, President John F. Kennedy committed the United States to "achieving the goal, before this decade is out, of landing a man on the Moon and returning him safely to Earth."

President Kennedy calls for a mission to send man to the moon during a joint session of Congress on May 25, 1961.

Upon hearing the President's promise, Thomas M. Moore, then a civilian radar expert for Dr. Wernher von Braun's team of rocket scientists, realized that the time had come for the perfection of a rocket-propelled Astronaut Maneuvering Unit (AMU), such as the one he'd originally conceived of during the late forties.

Dr. Wernher von Braun and his team of rocket scientists had arrived in the United States from Germany after the end of World War II. In support of America's new manned spaceflight program, they began experimenting with propulsion methods to get man up to outer space.

Dr. Wernher von Braun

How man might maneuver in outer space or on a lunar surface was a daydream that fascinated Thomas.

"As early as 1946 I began sketching out an Astronaut Maneuvering Unit Development Plan wherein I, in conjunction with Dr. von Braun's plans to put an astronaut in outer space, would design and develop an Astronaut's Maneuvering Unit for the astronaut's Extra Vehicular Activity (EVA)."

Believing that man, as President Kennedy later promised,  would walk on the surface of the moon one day, in 1949 Thomas presented von Braun with the first of many conceptual designs for AMUs.

Over the next decade his AMU designs would vary in size and shape and focus on military applications, but his first AMU design was for a small, light and portable earth- and space-use strap-on unit.

"In early 1952, with von Braun's support, I applied for and received $25,000 in military funding to begin small portable AMU experiments. Later that year the very first unit designed was born."

"Imagineering" is the word that comes to mind when fusing daydreams with spaceflight engineering, and if Thomas's daydreams were to come true, a small, light and portable AMU could conceivably help the military too.

"Wearing a modified AMU an ordinary soldier would have extraordinary abilities during air-to-ground movements, when combing through rough terrains, and when scaling difficult cliffs."

Thomas also pointed out that once fully developed, the most exciting application of his portable AMU was in man's exploration of the moon…and lessons learned through experiments on Earth could be applied in outer space such as ferrying astronauts from ship-to-ship, servicing orbiting spacecraft and surveying the surface of the moon.

"Von Braun was going to strap a man onto a rocket; I was going to strap a rocket onto a man."

Designing the first of his mobility units made Thomas feel as if he were part of a larger dream, the one shared by millions of kids, who like him, had cheered on Buck Rogers and Wilma Deering as the two comic book heroes soared across the silver screens. And more times than not his aides would call him Buck, as Thomas planned to be the first to try his luck.

The Jetvest in Science Fiction

Thomas's daydream of space flight seemed one-step closer to reality as the time had arrived to begin his AMU experiments.

"During the first week of indoor tests the program's AMU, a small chest worn device with a parachute harness, jet nozzles at shoulder level, and control knobs on the chest, was fitted onto a tethered dummy on a stand. Once the propellant (compressed air) was discharged, the unit lifted the dummy several feet into the air. Pitch, yaw and roll were still problems that lay ahead, but the unit was operational and I felt it was time to take the plunge myself."

With the help of his aides, Thomas built a 20 ft. high outdoor testing stand, donned white coveralls and steel helmet and took the unit outdoors for its first tethered human test.

3-2-1-   Whoosh

Thomas rocketed up 20 feet to the top of the testing stand, hovered a beat and then came back down slowly. What would he call his new AMU? As it was powered by jets and worn like a vest, it would be called "The Jetvest."

Tests were promising and von Braun was enthusiastic about the potential of the AMU program. However, the military faced other priorities and later that year further funding of the program ceased. After President Kennedy's speech, in 1961, spaceflight engineers began experiments towards the development of a Life-Supporting AMU, believing, as had Thomas, that future astronauts would need an AMU to maneuver about in zero gravity.

The Jetvest

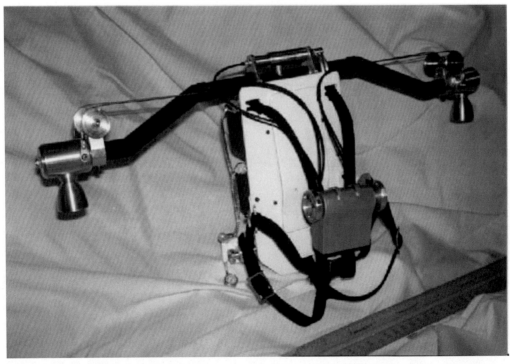

A model of Thomas Moore's Jetvest is on display at the U.S. Space and Rocket Center, Huntsville Alabama (courtesy Aerospace Engineer Mark Wells)

Thomas retired from the American Space Program but he never let his solo spaceflight daydreams die. When the opportunity to join Space Camp (a summer camp for kids who share similar spacewalking dreams) arrived, Thomas found a new passion. When asked in the early 1990's how he felt at being the first to turn science fiction into fact and even take a plunge at flying like Buck Rogers, Thomas was at first shy and reluctant to reply. Then thinking of the exhilarating WHOOSH of his first short tethered flight, Thomas sat at his desk and softly said "nice."

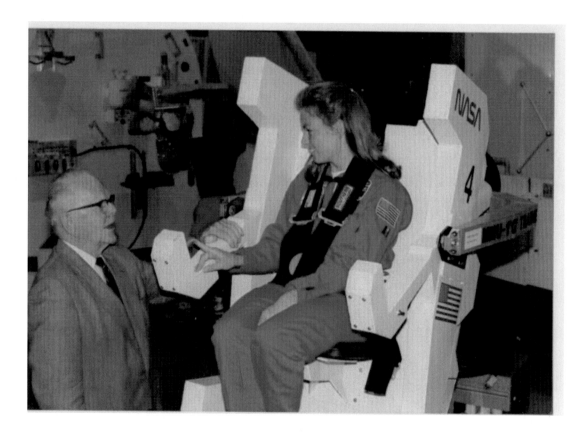

Thomas Moore and Assistant Program Manager
Karen Sirianno review The Space Camp MMU's operation.

Postscript:

To salute both the physicist and his Jetvest invention, in 1993, the U.S. Postal Service unveiled a space fantasy stamp that depicted two aeronauts soaring through the air with the aide of modern Jetvests. During the issue ceremony at the U.S. Space and Rocket Center in Huntsville, Alabama, Thomas was recognized as a spaceflight visionary and a man ahead of his time.

United States Postal Service Space Fantasy Jetvest Stamp

Physicist and Jetvest Inventor
Thomas M. Moore
U.S. Space Camp

The Jumpbelt

The story of the Jumpbelt and Flying Belt experiments begins not in the lab but during WWII at the West Point Military Academy. Lt. Charles Parkin was giving a class demonstration of flamethrowers powered by tanks of highly compressed nitrogen gas. While cleaning up, a soldier accidentally opened a valve on a tank of nitrogen gas and the force it released started Lt. Parkin speculating. He knew that opening the valve all the way would send the tank racing forward, like a toy balloon whooshing across a room. Looking at the tank Parkin had an idea: Could an infantryman hitch a ride?

After the war, Parkin, now a Colonel, returned to his idea. He strapped a tank of nitrogen gas on his back and found that it consistently boosted his jumping ability. Parkin, ideas in hand, decided to investigate what companies with rocket propulsion labs could develop by way of an accessory propulsion device. What Parkin discovered was that two engineers at the Thiokol Chemical Corporation were thinking along the very same lines.

Harry Burdett and Alexander Bohr, Thiokol's chief rocket engineers, were also interested in boosting an infantryman's abilities. For developmental purposes, they divided personal rocket propulsion devices into two categories: "Jumpbelts" and "Flying Belts." To assist a soldier over dangerous terrain they envisioned the Jumpbelt, a compact, lightweight, jet thrust unit worn around the waist, which would improve a soldier's running, jumping, and even enable water skimming.

On the other hand, to assist the soldier in flying over dangerous terrain, they envisioned the Jet Flying Belt. The Jet Flying Belt would be strapped on the soldiers back and with its jet engine would be capable of lifting the soldier into the air, permitting him to fly for several miles.

However, a miniature jet engine had yet to be invented so Harry and Alexander chose to experiment with their waist-worn thrust rocket unit first. It was called the Jumpbelt and given its capabilities the project became affectionately known as "Grasshopper."

In 1958, the Army's interest in Small Rocket Lift Devices (SRLD) brought Parkin to Thiokol to witness the first test of a Jumpbelt. Harry and Alexander explained that a 350-pound, five-second thrust from the Jumpbelt would

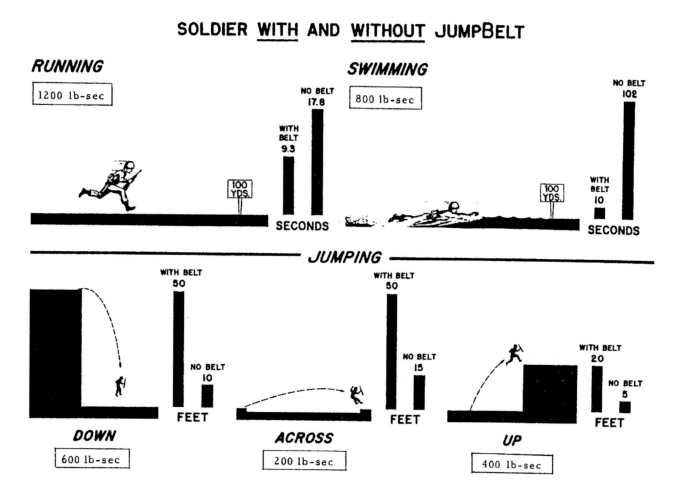

SOLDIER WITH AND WITHOUT JUMPBELT

provide the boost a soldier would need to jump 30 feet horizontally, leap to an altitude of 15 feet, and run at 22 miles per hour. Their preliminary tests were successful. Alexander believed that with continued funding and development, wearing an advanced Jumpbelt would make it possible for a soldier to run a hundred yards at 30 miles per hour, jump across 40-foot wide streams, or leap 20 feet onto a building or wall!

When asked what it was like to use a Jumpbelt, operators Ray Wiech and Stan Kedzierski described the experience of a jumpbelt thrust: "it was like taking off on a fast elevator, swinging in a parachute harness for a moment and then being gently lowered like a puff of milkweed."

First Look: The Hush-Hush Flying Belt

THESE hitherto unreleased photos show an early model of the secret "Flying Belt." In view at right, jets and control are folded for maximum convenience; in the other they are in flying position. Thrust angle is changed by moving the control forward or back, and amount of thrust or lift is varied by the movement of the hand grip.

The belt would enable the wearer to rise quickly to an altitude of several hundred feet and fly "miles."

How soon will the Flying Belt emerge from the laboratory? Reaction Motors' engineers estimate delivery of a workable one within two years.

Thinking back to his nitrogen gas tank experiment, Colonel Parkin wondered if a SRLD capable of sustaining flight could be constructed. With that concept in mind Harry and Alexander went back to the engineering lab and designed a rocketbelt which they called a "Flying Belt" named after the one used by Buck Rogers.

Later that year the Thiokol Corporation lost their bid for military funding to continue development on the Jumpbelt or a SRLD capable of flight and reassigned Harry and Alexander to other rocketry projects.

Though they never experimented beyond the Jumpbelt stage, Harry's and Alexander's experimental SRLD data led propulsion engineers one-step closer toward the day when man would fly free.

Alex Bohr and Harry Burdett Jr. examine five-canister, two-nozzle belt worn by Ray Wiech.

The Aeropak

While Harry's and Alexander's Jumpbelt brought about military interest, it was their Flyingbelt concept that the military wanted to further explore. In 1959, a research contract for a SRLD with flight capability was announced. The Aerojet General Corporation, a manufacturer of liquid and solid propellant rocket engines was awarded the contract.

Aerojet's engineers evaluated various designs including a stand-on pogo and sit-down chair SRLD before deciding on a strap-on Buck Rogers type of rocketbelt, which they called The Aeropak.

During the eight-month research & study period, project engineer George Trudeau and his staff focused on two vital aspects: (1) finding the most powerful yet safest fuel possible for the SRLD, and (2) combining stability with maneuverability and control. They chose hydrogen peroxide because it could be discharged easily, its flow could be regulated, and there would be no poisonous or harmful emissions.

Although the temperature within the rocket could reach a frightful 1370 degrees Fahrenheit, the nozzle's air temperature registered at a mere 100. Thus, there would be little likelihood of injuring people nearby. Should the noise in-flight be deafening, the operator could wear earplugs. Center of gravity, thrust capability, weight, the precise timing of ignition and control were the second considerations. To control pitch and longitudinal translation the Aeropak was given gimbaled nozzles; to control yaw - jetavators.

In theory, the calculations indicated that this Buck Roger's type flying unit would be stable and that even an unskilled person with limited training could safely operate it. Numerous indoor tethered tests followed which demonstrated that the Aeropack was capable of one-man short-distance flight.

Early in 1959, George tested their prototype Aeropak Rocketbelt in public at their Azuza proving grounds in California. The Aeropak's light metal tubing harness was fastened to a sheet metal back plate to which engineer Richard Peoples strapped himself into by means of chest and leg safety belts.

Aerotest pilot Richard Peoples
prepares to tether test the Aeropak.

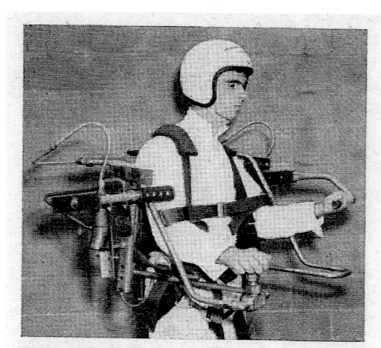

ROCKET WINGS. Aerojet-General's rocket pack is designed to let GI hover, fly vertically or horizontally. The left hand controls angle of thrust; the right, speed.

For the open-air test of the Aeropak, George chose a limited flight of 100 feet in distance and a maximum 30 feet in altitude. To avoid any serious mishap, the Aeropak was tethered by rope to a ground crew. For flight direction Richard regulated the angle of the two 300 pound thrust nozzles by using a lever in his left hand. For speed control (fuel flow) he used a right-hand valve/throttle.

An aura of excitement surrounded the engineers as Richard gave the throttle a slight twist and the very first rocketbelt roared. Then he gave the throttle a full blast; the sound was deafening! Richard rose up, moved forward then throttled down to land gracefully, completing the first successful outdoor tethered flight of a Buck Rogers type rocketbelt.

Have Aeropak will travel!

Proudly, George Trudeau envisioned Aeropaks not only for members of the military, but also for every member of the family.

Because of Aerojet General's successful rocketbelt research, the military solicited new bids for the second phase of their SRLD contract: a fully tested free-flying model.

While strolling leisurely one day with his friend and co-worker Jim Powel and discussing the SRLD military contract, Wendel F. Moore, a Bell Aerosystems Propulsion Engineer, reasoned that both a soldier and an astronaut could soar freely and untethered when aided by the controlled thrust of a small rocket.

Wendell had speculated on this topic as early as the 1950's as had von- Braun's Physicist Thomas Moore (no family relation). Unlike Thomas, Wendell had no opportunity to test his theories.

The military's1959 SRLD research and study contract motivated Wendell to request funding from Bell Aerosytems to do in-house research. Bell approved with the view that Wendell's exploratory work might win the military's study contract. Despite the data obtained in Wendell's first phase of experimentation however, Bell Aerosystems lost the SRLD contract to Aerojet General.

In August of 1960, the military solicited bids for the second phase of their SRLD study and awarded the contract to Bell Aerosystems.

Wendell F. Moore who was then the assistant chief at the Bell Aerosystems' Buffalo site was selected to build a non-tethered free-flying Small Rocket Lift Device (SRLD).

The dream of flying like Buck Rogers as shared by kids worldwide.

Harold "Pete" Graham: First to fly like Buck Rogers
(as recalled by Harold "Pete" Graham)

As a youngster during the '40's, nothing stirred my imagination more than reading about astronaut Buck Rogers, the 25th century earth and space hero. I would count the days till Saturday when I could visit the local Bijou theater and watch Buck and Wilma rocket ride across the screen. Then, not to be outdone, every other week, *Commando Cody* would jet across the sky as *King of the Rocketmen*! Free flight with the aid of a rocketpack as seen in the comics, movies and television has always been a dream. In 1961 I helped turn that dream into reality.

It was the dawn of the Space Age. Gagarin, Titov, Shepard and Glenn, were forging a pathway to the stars. At Bell Aerosystems in Buffalo, NY, a small team of rocket engineers headed by Wendell F. Moore, were paving the way towards mankind flying free.

Wendell had been part of the engineering team that worked on the X-1 rocket plane, which allowed test pilot Chuck Yeager to break the sound barrier in 1947. Now Moore was applying his knowledge and experience to constructing a rocket belt just like the one in the Buck Rogers comics.

Now the idea of individual rocketbelt propulsion was not new. During WWII German rocket scientists envisioned rocket-propelled combat soldiers.

However, it wasn't until Wendell F. Moore and his specialized rocket-team of engineers ironed out the kinks inherent in the man-rocket combination during the second phase of the program in 1961, that I would have the opportunity to fly unencumbered.

To gather operational data during the first few months of independent research and development at Bell, Wendell F. Moore and his close friend Bell Aerosystems' camera operator and cinematographer Tom Lennon, strapped themselves into test-rigs.

The data Wendell derived from those first experiments with himself on the rig established that man as a flying object was inherently unstable. From Tom Lennon's rig tests, Wendell learned that pilot attitude and physical balance played a key role in maintaining stability.

Aerotest Pilot & Bell cinematographer Tom Lennon

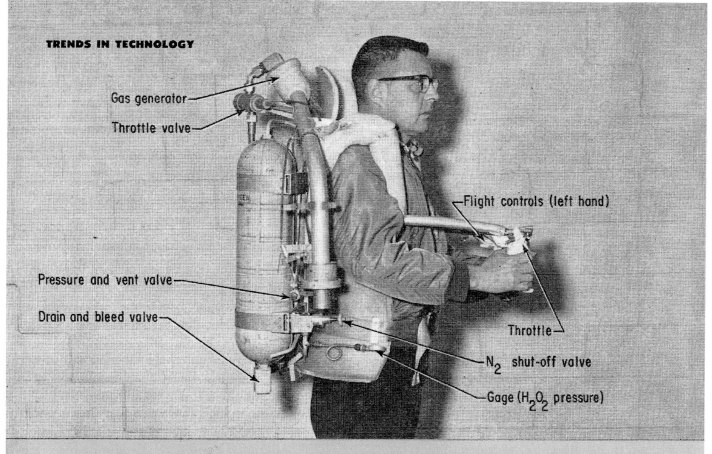

Gas generator

Throttle valve

Flight controls (left hand)

Pressure and vent valve

Drain and bleed valve

Throttle

N$_2$ shut-off valve

Gage (H$_2$O$_2$ pressure)

Present rocket-belt configuration is changed little from the feasibility model demonstrated to the Army in 1961. Engineers researching the belts say stability and control factors represent the most serious problem. Energy available from present fuels and man's ability to "back-pack" the fuel and machinery limit flights to 21 sec.

Circle 29-C on Page 19 for extra copy.

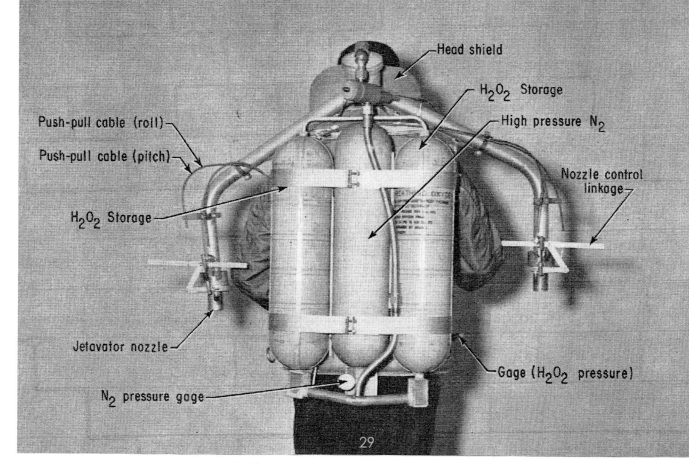

Head shield

H$_2$O$_2$ Storage

High pressure N$_2$

Push-pull cable (roll)

Push-pull cable (pitch)

Nozzle control linkage

H$_2$O$_2$ Storage

Jetavator nozzle

Gage (H$_2$O$_2$ pressure)

N$_2$ pressure gage

During the second phase, rocketbelt technician Ernie Kreutinger, built Bell's first experimental rocketbelt. One afternoon, while Wendell was personally testing it, simulating Superman's horizontal flying position, a cable snapped and he took a spill. An injury to his kneecap permanently grounded him.

The call for a permanent pilot sent images of Buck Rogers and Commando Cody racing through my mind. I'd don a helmet, strap on my rocket pack, blast those venom-spewing lizard-men from the distant planet X, rescue the lovely princess Kathleen and show that Flash Gordon guy a thing or three!

I was working at Bell Aerosystems, in my twenties, 6ft tall, an ice hockey player and in perfect physical shape. Wendell looked me over, said "yes" and I was on my way to becoming the first to fly a rocket belt, a real life Rocket Ranger!

Gradually I familiarized myself with this unpredictable and volatile machine. Rocket-riding handbooks did not exist and I couldn't phone Buck Rogers for his assist.

The team and I created flight procedures daily and mostly through trial and error and great risk. Preparing for any eventuality Wendell Moore assembled a ground crew of true rocket rangers: firemen Norm Sherry and Mitch Crevar, engineers Eddie Ganczak, Ernie Kreutinger Leo Thompson and John Kroll. Support members Glen Pabst, Tom Lennon, Marve Deboy and nurse Millie George.

The Bell Rocketbelt Team

My first tethered flights inside a Bell aircraft hanger were an assortment of erratic bobbing and weaving motions. It took 36 tethered flights to complete the debugging, lift-off, flight and landing procedures.

Wendell's bucking miniature rocket had finally been tamed! For comfort during those 36 tests, pads and supports were placed under my arms and at my waist and abdominal area. Though technically we were ready for flight, the rocket belt was still an experimental flying machine in its rawest form. I had no way of knowing if I was going to actually fly with this contraption on...or fall, crash, and burn!

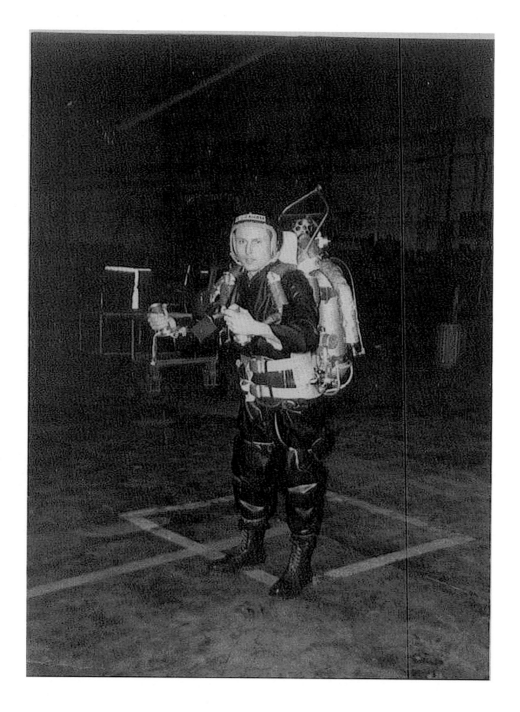

Harold "Pete" Graham finishes his training.

To mount the rocket belt, I would slip my arms through padded lift rings and then secure the unit with

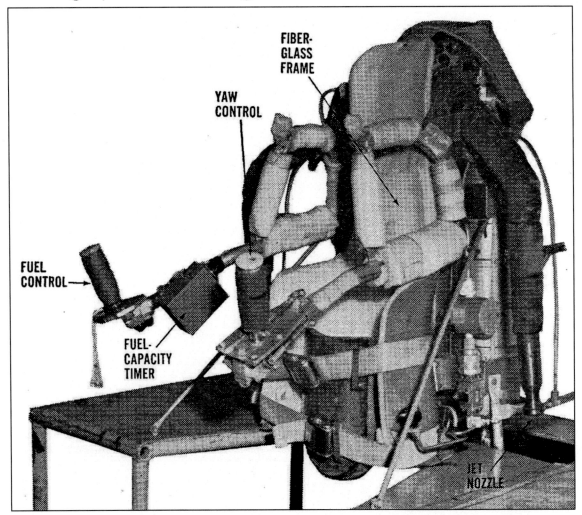

FIBER-
GLASS
FRAME

YAW
CONTROL

FUEL
CONTROL

FUEL-
CAPACITY
TIMER

JET
NOZZLE

two quick-release belts around my abdomen. It had two metal tubes attached to the lift rings extending forward under my arms on each side. A control stick on the left tube permitted me to change flight direction.

A motorcycle-type hand throttle on the right allowed me to regulate rocket thrust levels, thus controlling my rate of climb. A throttle propulsion system forced pressurized hydrogen peroxide into a gas generator, where it mixed with a silver catalyst and decomposed into steam. As the steam escaped through the rocket nozzles, it accelerated to 300 pounds of thrust directed towards the ground.

The amount of fuel the belt could carry however only allowed for 21.5 seconds of actual flight time. Anticipating the many distractions that I might face, and the possibility that I might run out of fuel in mid flight, the engineers designed a vibrating fuel indicator. Placed inside my helmet, it would vibrate when the fuel was low, signaling me to begin my descent. Though Wendell agreed to protective clothing similar to Buck's and Cody's (a black rubber suit, industrial boots, a helmet, and goggles), he nixed my Hans Solo-type holster and silver lizard-men blaster.

The day arrived, April 20th, 1961, on a taxiway at Niagara Falls International Airport. Having trained for months on a tether, I was ready to meet the challenge of flying free: the successful operation, flight and landing of the first rocket-powered maneuvering unit. But this flight would be different; it would be untethered. Millie George, our Rocket Team nurse, characterized me as "office pale and slide-rule slender," referring to Clark Kent, Superman's alter ego in the comic strips. The time had come to sling on my twin jet cloak and give life to my comic book heroes.

The rocket belt team had prepared for any eventuality and stood vigilant. A quick test blast, a deep breath, and I was ready. With a full-turn of the throttle, the rockets strapped to my back roared ferociously. The cold temperature turned my heated exhaust into a white steam–like fog. I couldn't even see my feet! Rising slowly, I pressed down on the control arms, which thrust me forward.

Rocket Ranger Harold "Pete" Graham
April 20th, 1961

I hovered forward slowly until I landed safely at my touchdown point one hundred and twelve feet away, eight feet less than the Wright Brothers had flown with wings at Kitty Hawk. The flight captured the public's imagination and coincided with the first flight of man in space.

Flying All Over The World!

For over a year I traveled the world demonstrating Bell's rocketbelt and its potential future use by civilians, soldiers and most importantly by astronauts. The need for man to maneuver about in space or on a planetary body was becoming increasingly clear, as humankind's dream of space travel became a reality. In those early days of testing, we could only imagine what it would be like to one-day reach for the final frontier. So popular had the thought of Flying like Buck Rogers become that I was invited to do appearances on TV. My favorite appearance: To Tell The Truth (1962) with Johnny Carson and Betty White.

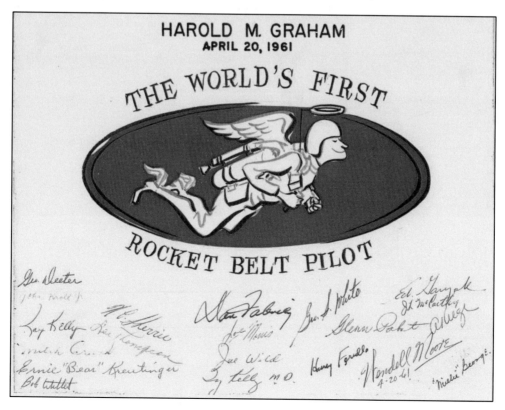

My most memorable demonstration flight was for President Kennedy at Fort Bragg in October 1961: rocketing 200 feet away from an offshore amphibious vehicle, I landed in front of him, stood at attention and saluted. With his vibrant red hair, he stood out in the crowd; he looked amazed, wide-eyed and open-mouthed, just like a kid!

Harold Graham flies in for a landing in front of J.F.K.

Harold "Pete" Graham salutes President John F. Kennedy

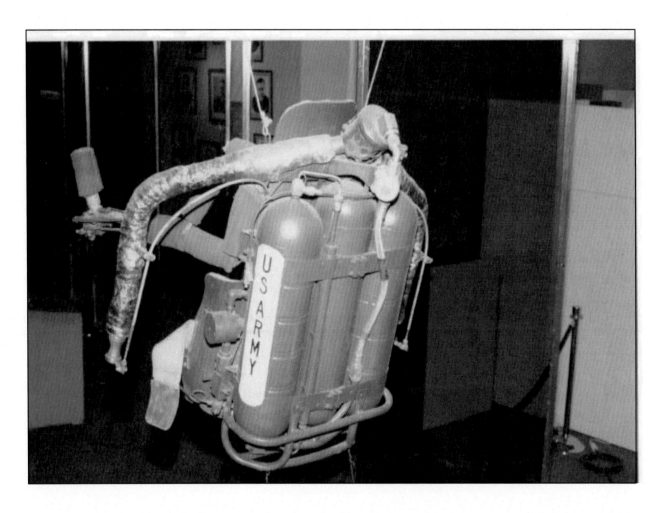

Harold "Pete" Graham's Rocketbelt can be seen on display at
The US Army Transportation Museum in Fort Ustis, VA

The Bell Rocketbelt Pilots

 Entertainment promoter Clyde Baldschun was watching over his little niece when she saw me doing a demonstration flight on TV. Her amazed reaction was such that the next day Clyde called Bell and so began his management of the Bell Rocketbelt flight demonstrations.

To pursue my education I retired from the rocketbelt program in 1962. But with Clyde receiving inquiries for rocketbelt flight demonstrations daily, a team of Rocketbelt Pilots was assembled and trained. The flight team consisted of former Jet Pilot Robert Courter as team captain and rocketbelt pilots Peter Kedzierski, Gordon Yaeger and William Suitor.

For the adults the Pilots became Ambassadors of American ingenuity and for their wide-eyed cheering kids, a Buck Rogers come to life.

Robert Courter, the rocketbelt pilot's team Captain and rocketbelt engineer Wendell F. Moore.

By 1964 the Rocketbelt Pilots had given demonstration flights across the country and across the world.

Buck Rogers Goes to Paris

Counterparts of Buck Rogers will be at the International Aviation and Space Exposition, Paris, June 7-16, where a real rocket belt will be demonstrated.

Called the small rocket lift device by the Army, the current rocket belt was not designed to meet specific military requirements.

But, the SRLD is being studied by both the developer—Bell Aerosystems Co., part of the Textron Inc. complex—and the Army to develop it to be used to transport foot soldiers over surface obstacles and to permit assault troops to fly from ship to shore.

If that is not Buck Rogers enough, Bell is considering the use of the belt principle as a means of propulsion in weightless space environment. On the moon, which has one-sixth of the earth's gravity, rocket belt performance, the company said, would be increased five-fold.

While research and development efforts at Bell are aimed at reducing the weight of the rocket belt and increasing its range, the current system has propelled its operators in controlled free-flight over ground distances of 815 feet at 60 miles an hour and heights of almost 60 feet.

Wendell F. Moore, the Bell engineer who invented the belt, which consists of a twin-jet hydrogen peroxide propulsion system mounted on a fiberglass corset, described the device as "a feasibility model designed to prove that lightweight rocket power can lift a man and transport him over the ground in controlled flight."

A control stick on one tube permits the operator to change his direction and a motorcycle-type handle throttle on the other tube allows him to regulate thrust levels—control of his rate of climb and descent.

Peter L. Kedzierski, one of the two operators who will perform at the IASE said, when flying in the belt "it feels like a giant is picking you up under your arms."

Rocketeer Peter Kedzierski greets admirers after his flight

However, nothing could compare with their demonstration flights at the 1964-5 World's Fair in Flushing Meadows Queens, New York, where flying as Kolonel Keds, the hero of the Keds sneaker company, they received the thunderous cheers of thousands as they daily circled the emblem of the world's fair, the colossal

Unisphere Globe!

Bell Aerospace TEXTRON   Buffalo, New York 14240   .198009.

Rocketeer Robert Courter smiles for the camera after his
Unisphere Globe flight

An official World's Fair Kolonel Keds comic book

The "Belt" as worn by the original Bell Rocketbelt pilots
is on display at the National Air and Space Smithsonian's
Steven F. Udvar-Hazy Center

And then came TV!

Following in my footsteps the Rocketbelt Pilots appeared on the most popular television programs at the time from Gilligan's Island to Lost in Space. And even late night host Johnny Carson gave Robert Courter's rocketbelt a go. But the biggest thrill for all came when Pilots Yaeger and Suitor doubled for James Bond in the secret agent's thrilling *Thunderball*.

To apply all his time and resources to the creation of a new Jet Flying Belt design, in 1967 Wendell Moore discontinued rocketbelt flight demonstrations. Then in 1969 in association with jet engine manufacturer Williams International, Wendel F. Moore unveiled the successor to his rocketbelt, the first ever Jet Flying Belt. To everyone's sad disbelief, on the day of his Jet Flying Belt's first flight Wendell Moore died. At that time he was also developing designs for rocket-powered AMU's that could be used on the moon and in zero-gravity.

The Tyler Rocketbelt

The idea of rocketbelt flight demonstrations didn't end with the cancellation of Bell's demonstrations program. Thinking of one day flying a rocketbelt himself, California-based camera systems inventor, Nelson Tyler studied Bell's rocketbelt's design and then built his own modern version, the Tyler Rocketbelt.

Nelson Tyler prepares his rocketbelt for flight

Now building a rocketbelt wasn't just an inventor's whim for Nelson; he had wanted to fly like Buck Rogers ever since he was a kid and listened to Buck's show on the family radio. With rocketbelt pilot Bill Suitor's training, in the early 1970's Nelson realized his childhood dream.

Nelson Tyler goes flying like Buck Rogers

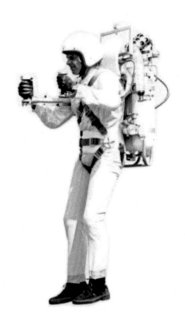

Throughout the mid 1970's, Bill Suitor kept the Tyler rocketbelt in flight in commercials and on TV but inventive demands on Nelson's time didn't allow for managing rocketbelt flights and in the late 70's he retired his rocketbelt and stored it away.

Nelson Louis Olivo with the Tyler Rocketbelt

In 1980 with the financial support of Nelson Louis Olivo, the guidance of original 1960's rocketbelt demonstration promoter Clyde Baldshun, and the flight expertise of former Bell Aerosystems rocketbelt Pilot Peter Kedzierski, Tyler's rocketbelt was resurrected and prepped for a new era of demonstration flights.

However, it wasn't until 1984 that the most important call for an Ambassador of American ingenuity and a real life Rocket Ranger arrived. In 1980, Olympic pyrotechnics Director Tommy Walker, Nelson Louis Olivo, and promoter Clyde Baldshun agreed on an opening ceremonies rocketbelt flight for the 23rd Olympiad; and now the time had arrived. But, with flight insurance and rocketbelt fuel both very difficult to acquire, could an Olympic Rocket Ranger be provided in time?

Bill Suitor and the 23rd Olympiad

With 2 billion people across the world watching on their television sets, Bill Suitor, the last and youngest of the original Bell Rocketbelt pilots and the personification of a real Rocket Ranger, gave a twist to the familiar Tyler Rocketbelt's throttle and WHOOSH lifted up and into the air as part of the spectacular opening ceremonies at the 23rd Olympiad!

Like a lone dove, Rocketbelt Pilot Bill Suitor glides down and across the arena of the Los Angeles Coliseum capturing the hearts of the young and the imagination of the world.

Over mesmerized eyes Rocketbelt Pilot Bill Suitor begins his descent.

Spotting his landing grid Rocketbelt Pilot Bill Suitor eases down to a smooth two-point landing

To Nelson and John..... heres' to a great
book and a great future
Bill Suitor

Rocketeer William "Bill" Suitor

The worldwide interest that followed Bill Suitor's Olympic flight gave birth to a new generation of young rocket rangers whose rocketbelt flights to this very day continue to inspire new and wide-eyed awes.

Every so often, I am asked to reminisce about my early days of rocketbelt flights and the pioneering efforts of engineer and inventor Wendell F. Moore. And whenever I am, I feel inwardly proud. Though our contributions towards maneuvering through space were minimal, Wendell F. Moore, his ground crew and the Bell Aerosystems team of Rocketbelt Pilots not only turned science fiction into fact, they helped to forge a path towards one-day humankind flying free through air and space.

To celebrate the 30th anniversary of the first rocketbelt flight Harold Graham's family and friends, surviving original rocketbelt ground crew members and Nelson Louis Olivo (presenting Harold with a Real Rocket Ranger Plaque), gathered in Buffalo N.Y.

Harold Graham holds up his Rocket Ranger plaque while Millie George, the original Rocketbelt Team nurse, and invited guests await his acceptance speech.

Rocketbelt Wings
(as recalled by Rocketeer Gordon Yaeger)

   In looking back at my days as a rocketbelt pilot, the most important aspect of my experience was my flight training. The rocketbelt was an incredible blend of the colossal and the diminutive: a miniature version of an Atlas rocket. And though it appeared rather easy to fly, behind the scenes a potential rocketbelt pilot trained very hard to earn his rocketbelt wings. Bell Aerosystems Rocketbelt pilot training began with a ground school instrument course wherein the pilot became familiar with the placement, feel, weight, and operation of the control handles. Then a pre-flight check allowed the rocketbelt pilot and the rocket fuel techs to inspect, check and fuel the rocketbelt before liftoff.

With the rocketbelt on and safely tethered, the trainee Rocketbelt pilot was required to complete 50 training flights designed to develop flight, control and maneuvering techniques. After fifty tethered training flights were successfully completed, the rocketbelt was untethered, inspected and refueled for the trainee pilot's solo.

Gordon Yaeger in tethered training

The Pilot Solo

The Pilot:

1.   Straps on his rocketbelt
2.   Checks his rocketbelt's gages.
3.   Unlocks the throttle handle,
4.   Moves the control handles to the neutral position
5.   (10 degrees forward, 0 degrees aft, 0 degrees lateral), with the yaw handle/rudder centered.
6.   Then after taking a deep breath and giving a "ready" thumbs up, the rocketbelt pilot opens the throttle and WHOOSH!

Forward flight:

This requires a slight downward pressure of the control arms from the neutral position.

Backward flight:

The pilot relaxes the downward pressure of the control arms. There is a built-in upward force (approx. 20 lbs) on the control arms that gives the pilot a positive feel at all times.

Right or Left Turn:

There are two types of turns that a pilot can accomplish in a rocketbelt. A forward left- coordinate turn is made by slightly raising the right shoulder and turning the right yaw control handle left, (this blends the yaw and lateral shoulder movement of the thrust tubes, the rudder, and aileron). A right-coordinate turn is accomplished by performing the opposite maneuver.

A Hovering Turn (or full spin):

Turning the yaw control handle while hovering turns or spins the pilot's body left or right.

Landing:

Descent involves a combination of visual cues and throttle controls. The trainee pilot begins by cutting back slowly on the throttle, discontinuing any body movements, sighting the landing point and gently landing on mother earth.

Once the solo was successfully completed, the trainee pilot was now among the very proud few who had earned his Rocketbelt wings.

Gordon Yaeger coming in for a perfect two-point landing

Rocketeer Gordon Yaeger

The Bell Flying Belt (JetBelt)

(as recalled by AeroTest Pilot Robert Courter)

On April 7th, 1969 engineer Wendell F. Moore unveiled his new Bell Flying Belt, a jet-powered 20th Century version of Buck Roger's 25th Century mode of transportation and the successor to his world-famous Rocketbelt. Flown for the first time in public, it was sheer coincidence that the Jetbelt's maiden voyage was made at Niagra Falls, the capital city of Earth in the Buck Rogers adventures. Originally designed for military applications, the Jetbelt was also seen as a potential 21st century-type commuter vehicle.

Utilizing the same operational design and flight control mechanisms as the rocketbelt, the Jetbelt had only two significant differences: (1) a different, less volatile fuel (JP-4) and (2) a miniature fan-jet engine, the WR-19. Unlike the rocketbelt whose flights were measured in seconds and feet, the Jetbelt's flights were measured in minutes and miles.

INTRODUCING....
THE NEWEST DEVELOPMENT IN
PERSONAL FLIGHT

THE BELL
FLYING BELT*

* Patented
U.S. Pat. Office

As Captain of the world-touring Bell Rocketbelt Team, I had completed thousands of demonstration flights and this experience led to my being chosen to fly the Jetbelt. In flying the Jetbelt, I demonstrated how man might one day maneuver about, not in outer space, but in inner space.

I've often been asked what did a solo Jetbelt flight feel like? Well, it's "like having a giant pick you up by the arms," as Peter Kedzierski, one of the original Bell Rocketbelt pilots put it. Once I was up, there was a sense of freedom that reminded me of skin diving. It may seem odd, but I had a comfortable feeling of security aloft with the Jetbelt.

I actually felt safer up there than I did driving the family car. To fly with the JetBelt required a minimum of fifty tethered flights and once successfully completed, a pilot could then fly solo.

A flight with the Jetbelt begins with preflight, fuel, instrument and radio checks. When all systems are go, the pilot backs into the corset and straps himself in.

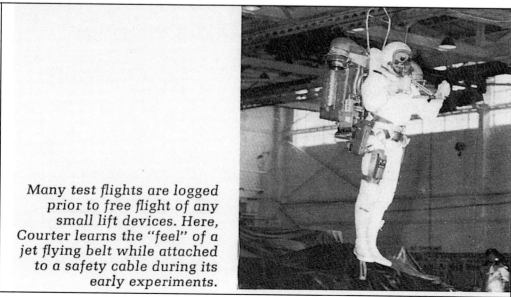

Many test flights are logged prior to free flight of any small lift devices. Here, Courter learns the "feel" of a jet flying belt while attached to a safety cable during its early experiments.

men of *action*

VOLUME 10, NO. 4 – 1970

MAN
WITHOUT
WINGS

page 10

While doing this, a retractable stand supports the machine on the ground. To start the miniature engine, the pilot triggers a solid-propellant cartridge that spins and ignites the fuel. With the throttle (a twist grip in the right hand) set at the "ground idle" position, the pilot checks controls and instruments. Then he advances the throttle to "flight idle," and the lift just offsets the weight of the pilot and belt, making it easy to move about on the ground. Before the pilot lifts off, he or she rechecks the instruments and retracts the stand. Then the pilot eases the throttle up, the r-p-m-s increase and the pilot is airborne.

After a year of demonstrations, television commercials, and magazine articles the Jetbelt was retired in favor of promoting the Williams International X-Jet, a wingless walk-on miniature fan-jet flying platform. Although the Jetbelt was never fully developed into that 25th century-type commuter vehicle, it allowed us all a glimpse into the world of the future, the world of Buck Rogers.

Courter, now with Williams Research Corp., firm that developed jet belt's small engine, congratulates Wendell F. Moore, inventor of the belt. Moore died unexpectedly in 1969 after many years of pioneering individual flying systems.

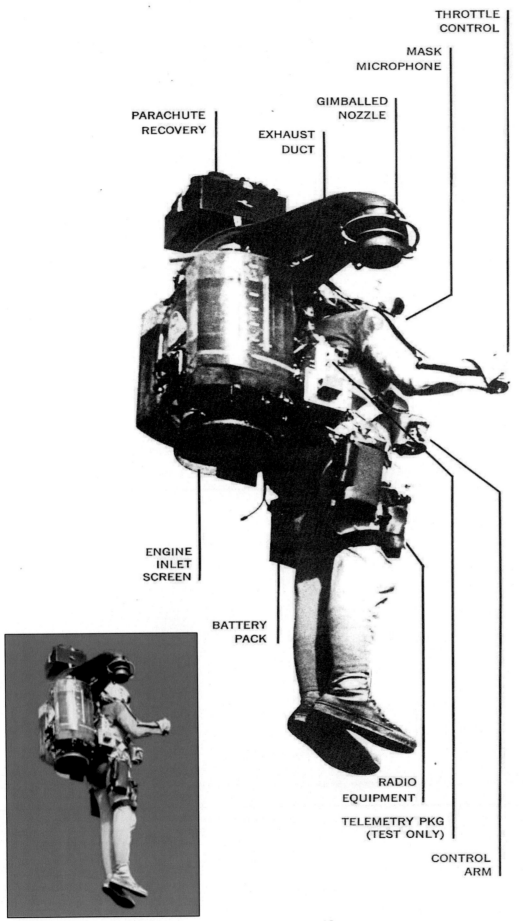

THROTTLE
CONTROL

MASK
MICROPHONE

GIMBALLED
NOZZLE

PARACHUTE
RECOVERY

EXHAUST
DUCT

ENGINE
INLET
SCREEN

BATTERY
PACK

RADIO
EQUIPMENT

TELEMETRY PKG
(TEST ONLY)

CONTROL
ARM

Best Wishes
Bob Courter

By 1930, Buck's sci-fi world of the future had captivated millions and while his devoted Rocket Ranger Astronauts dreamt they'd one day fly like Buck Rogers, adults would never believe that rocketships much less rocketmen, could or would ever exist outside of the monthly comics.

Were they right?

Volume 2 continues the quest!

Rocket Rangers Vol. 2

Thank you, Hal

Aerospace Historian Nelson Louis Olivo and Rocket Ranger
Harold "Pete" Graham pose together after completing
the first draft of this book (Dec.1991).

Printed in the United States
By Bookmasters

"Some books retell the history.
This book relives it."

—Thomas M. Moore, former physicist for
Dr. Wernher von Braun at Redstone Arsenal,
Huntsville, Alabama

ISBN: 978-1-4628-6546-8

90000>

9 781462 865468   (98417)

MAGNETICALLY INDUCED ELECTROMOTIVE FORCE

Frederick J. Young, Doctor of Philosophy
800 Minard Run Road, Bradford, Pennsylvania, US
15 October 2012

Michael Faraday, the Father of Emf